# Our Nature Chart

by Nan Walker

We take a walk.
We look and see.
We see yellow.

We see 1 yellow butterfly.
We see 2 yellow birds.
Look at our chart.

We take a walk.
We look and see.
We see red.

We see 2 red ladybugs.
We see 3 red ants.
Look at our chart.

We take a walk.
We look and see.
We see green.

We see 1 green snake.
We see 3 green frogs.
Look at our chart.

We see yellow.
We see red.
We see green.

# Look at our chart!